THE BEAUTY OF AURORA BOREALIS

I0440092

A Guide into the Mysteries of the Northern Lights

KATE OTTIS

COPYRIGHT © [2023] [KATE OTTIS]

ABOUT THE AUTHOR

Kate Ottis is a passionate traveler and author, celebrated for her captivating travel guides that inspire wanderlust and exploration. In her late thirties, Kate has carved a niche for herself by blending her love for adventure with her talent for storytelling. Her works often focus on uncovering the hidden gems of the world, offering readers practical tips and cultural insights to enrich their journeys. One of her notable works, The Beauty of Aurora Borealis: A Guide into the Mysteries of the Northern Lights, takes readers on a celestial journey to witness the awe-inspiring Northern Lights. This guide not only delves into the science and folklore behind this natural phenomenon but also provides practical advice on the best times and locations to experience it.

Kate's writing is characterized by its vivid descriptions, cultural depth, and a genuine passion for exploration. She aims to make travel accessible and meaningful, encouraging her readers to embrace the beauty and diversity of the world. Whether you're an experienced traveler or a curious beginner, Kate Ottis's guides are a treasure trove of inspiration and knowledge.

Table of Contents

INTRODUCTION

Dancing Lights in the Northern Heavens

The Aurora Borealis, also referred to as the Northern Lights, is one of nature's most mesmerizing and elusive displays that is frequently shown in the pitch-black darkness of the Arctic night. Humanity has been enthralled by these ethereal, otherworldly ribbons of light for generations as they dance over the heavens in a captivating exhibition. Nothing compares to the breathtaking splendor of the Aurora Borealis except for its mysterious nature.

Imagine yourself away from the bustle of the city, in the highlands of Iceland, the secluded wilderness of Alaska, or the Arctic tundra of Norway. There's a deep silence and a coldness in the air. The night sky starts to change as you look up, displaying an amazing show of colors that words cannot describe. Over the cosmic canvas, a shimmering green curtain with pink and purple undertones opens. It appears as though the gods are using their own divine brushstrokes to paint the sky.

The collision of charged particles with gasses in Earth's upper atmosphere results in a celestial dance

known as the Northern Lights. The interaction between Earth's magnetic field and solar wind is the remarkable scientific explanation. But knowing the workings doesn't take away from the feeling of mystery and wonder these lights arouse. We explore the core of this heavenly wonder in this book as we travel through the domains of science, history, culture, and firsthand experience. We'll look at the many tales and legends that the Northern Lights have influenced throughout history, learn how and why they appear, and discover the numerous ways they have impacted our planet.

Beyond scientific explanation, the Northern Lights are incredibly alluring. These lights have woven a global cultural tapestry for millennia, being worshipped as heavenly deities, fate-telling symbols, or even the souls of ancestors. We'll take you on a fast-paced tour of the many diverse cultural interpretations and myths surrounding the Northern Lights, ranging from Indigenous peoples' spiritual links to the lights to Viking beliefs in the Valkyries. We'll expose you to people whose lives have been significantly impacted by the Northern Lights, going beyond the facts and the mythology. Their individual tales, ranging from magical evenings spent beneath. the aurora to scientific research trips to the Arctic, offer a human perspective on this heavenly occurrence. However, the Northern Lights are more than just a breathtaking sight; they also represent a significant scientific mystery. The dance of the auroras helps us understand solar activity, Earth's magnetic field, and the broader forces shaping our planet's place in the universe. You'll understand the significance of this celestial spectacle even more as we work to solve the scientific puzzles.

This book is your ticket to a world of wonder and discovery, whether you're an amateur astronomer hoping to increase your chances of seeing this cosmic ballet, a tourist looking for the finest places to see the lights, or just someone who's fascinated by the idea of the Aurora Borealis. Get ready to be taken to a world where magic and science collide and the wonder and majesty of the cosmos take center stage as we set off on this adventure through the Northern Lights. As the reader, you are about to get a front-row ticket to one of the best exhibitions on Earth—the Northern Lights, nature's majestic spectacle

CHAPTER 1

Scientific Explanations of the Northern Lights

For millennia, people have been enthralled with the Northern Lights, also known scientifically as the Aurora Borealis in the Northern Hemisphere and the Aurora Australis in the Southern Hemisphere. Although they give the impression of being an amazing show of glistening, multicolored lights in the night sky, they are actually the result of intricate interactions between the solar wind, the upper atmosphere and the magnet of the sphere.

Charged particles and Solar wind:

The Sun, which is the source of the Northern Lights, is always releasing a solar wind—a stream of charged particles made mostly of protons and electrons. Particles and energy are brought into our solar system by this solar wind. A magnetic field is carried by the solar wind when it reaches the earth.

Magnetic Field of Earth:

Earth has an invisible shield-like magnetic field that emanates from its core and envelops the planet. The North Magnetic Pole and the South Magnetic Pole, which make up this magnetic field, are not always parallel to the North and South Poles in space.

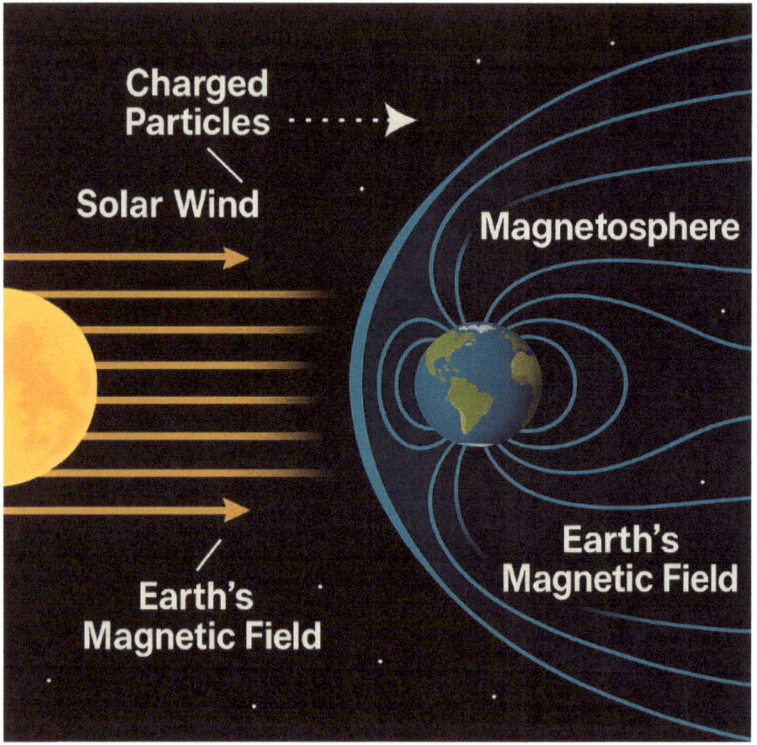

Reaction with the Magnetosphere of Earth:

The solar wind interacts with Earth's magnetic field as it gets closer, causing a phenomenon called magnetic reconnection. The charged particles from the solar wind can reach Earth's magnetosphere through gaps or mergers between the magnetic field lines of the solar wind and earth.

Auroral Ovaland Energy Transfer:

Energy is carried by the charged particles from the solar wind and is transmitted into the magnetosphere of Earth as they enter. A portion of these charged particles are directed toward the polar regions by Earth's magnetic field lines. As a result, the magnetic poles are surrounded by an area known as the auroral oval, which designates the best place to see the Northern Lights.

Excitation of Gases in the Upper Atmosphere:

These charged particles, mostly electrons, acquire energy as they move in the direction of the poles along the magnetic field lines. A portion of this energy is transferred to the atoms and molecules in the atmosphere when they collision with gases found in Earth's upper atmosphere, such as nitrogen and oxygen. Then, these energetic particles transition into excited or higher energy states.

Light Emission:

The extra energy is released by the excited particles as photons as they settle back into their lower energy states. The stunning hues connected to the Northern Lights are created by these light particles. Green and red hues are usually produced by oxygen emissions, whereas purple, blue, and pink hues are produced by nitrogen emissions.

Colors and Atmosphere Conditions:

The type of gas (nitrogen or oxygen) being excited, the height at which the excitation occurs, and the precise energy levels involved all affect the colors and patterns of the Northern Lights. The distinct hues and variations of the Northern Lights are caused by various gasses at various altitudes.

Atmospheric Layers and Altitude:

The precise hues and patterns of the Aurora Borealis depend on the height at which charged particles smash with atmospheric gases. The red and green tints are caused by oxygen emissions, which usually occur at higher altitudes (over 100 kilometers or 62 miles). Purples, blues, and pinks are caused by (between 80 and 100kilometers or 50 to 62miles).

Auroral Intensity and Solar Activity:

There is a strong correlation between solar activity, specifically the Sun's 11-year solar cycle, and the frequency and intensity of the Northern Lights. The Sun discharges more charged particles into space during solar maximum periods, which are times of increased solar activity and consequently more frequent and powerful auroral displays. On the other hand, the Northern Lights are fewer and more muted during solar minimum.

The Auroral Seasons and Earth's Tilt:

The tilt of Earth's axis affects how visible the Northern Lights are as well. Due to this tilt, there are occasionally possibilities to see the lights around the auroral oval, which is typically easier to see near the Antarctic Circle in the winter and near the Arctic Circle in the winter in the southern hemisphere.

Planets with Auroras:

While auroras resembling the Northern Lights are mainly seen on Earth, they can also be seen on other planets with magnetic fields. For example, auroras can be found on Jupiter, Saturn, and even Mars; these variations are caused by the distinct characteristics of their individual magnetic field and force.

Functioning Space Weather:

In addition to being a fascinating natural phenomenon, auroras are also quite important in the larger scheme of space weather. Comprehending the principles underlying the Northern Lights is crucial for researching space weather and its effects on satellites, communication networks, and even Earth's power grids.

Continued Studies and Space Travel:

Researchers never stop examining the Northern Lights in an effort to learn more about Earth's magnetosphere and solar-terrestrial interactions. Numerous space missions have been launched to collect data on the Northern Lights and their effects on Earth's ecosystem, including NASA's THEMIS and ESA's Swarm. Evidence of the complex dance of cosmic forces that shapes our globe can be found in the science underlying the Northern Lights. These lights are actually a dynamic and constantly-changing representation of our planet's interaction with the immense universe beyond, despite their seemingly peaceful and celestial ballet appearance. You'll come to recognize the Northern Lights as a natural phenomenon that provides the answers to comprehending the cosmic forces that control our planet as we go deeper into the scientific complexity.

CHAPTER 2

A History of Human Interest in the Northern Lights:

For many generations, the Aurora Borealis, often known as the Northern Lights, has enchanted people with its bright presence. People from many civilizations around the world have been captivated by these celestial events, and this has had a lasting impact on their mythologies, histories, and ways of life. Examining the historical background of the Northern Lights enables us to recognize the timeless and all- encompassing magic these lights inspire.

Legends and Old Beliefs:

The Northern Lights were a source of wonder and mystique for numerous ancient cultures. According to Norse mythology, the lights represented the shieldmaidens Valkyries, who selected soldiers deserving of entry into Valhalla, and their armor. The Northern Lights were a source of wonder and a potent omen for the Vikings, who regarded them as a manifestation of the gods. The lights had great spiritual significance for the Inuit and Sami, two traditional civilizations living in the Arctic. They were perceived as the departed's spirits,

Cultural Interpretations:

Dancing over the heavens and escorting their souls to the hereafter. In addition to being a beautiful sight, the Northern Lights served as a symbol of life's circle and a link to their ancestors for these people.

Enlightenment in Science:

Different cultural interpretations of the Northern Lights emerged as humanity spread throughout the world. The lights were known as "banners" or "flames in the sky" in ancient Chinese writings, and their appearance was frequently interpreted as a portent of momentous occasions, both fortunate and unfortunate. In a similar vein, Northern Lights stories were woven into oral traditions and creation myths by Native American groups such as the Algonquin and Cree. A lot of Russian folklore describes the "northern dawn," and the lights are frequently represented as fire-breathing animals or as the flames of woodland spirits. The lights are known as "revontulet," or "fox fires," in Finland. According to folklore, a fox creates them by sweeping its tail across the snow and shooting sparks skyward.

Interpretations of the Northern Lights changed along with human knowledge of the natural world. Scientists like Edmond Halley and Anders Celsius tried to provide a better logical explanation for the occurrence during the European Enlightenment. But even as science gained more insight, the mystery surrounding the Northern Lights persisted.

The goal of trips to the Arctic and Antarctic in the 19th century was to observe and record the Northern Lights. Legendary explorers and scientists like Fridtjof Nansen and Roald Amundsen braved these frigid environments to learn more about the nature and because of the lights while also taking in their majesty.

Object of Modern Reverence:

The Northern Lights are still a source of inspiration and awe in modern times. Adventurers from all over the world set out for high latitude locations to see this amazing show. The Northern Lights are now recognized as a symbol of pride and identity in nations like Iceland, Canada, Norway, and Iceland in addition to being a natural phenomenon. Creative inspiration might also come from the Northern Lights. Numerous literary, musical, and artistic works have featured them. They represent mystery, beauty, and the unbreakable bond that unites all living things with the universe.

Representations in Culture and Art:

In addition to influencing many cultures' stories and beliefs, the Northern Lights have served as an inspiration to innumerable singers, authors, and painters. Northern Lights have been depicted in works of art and literature ranging from Robert W. Service's poem "The Shooting of Dan McGrew" to Vincent van Gogh's "Starry Night." The Northern Lights have long been honored in Saami and Finnish cultures via music, dance, and narrative. The bright designs of traditional clothing, fabrics, and pottery reflect the vivid hues of the lights. These creative interpretations attest to the auroras' continuing culture relevance.

Scientific Progress and Cross-cultural Cooperation:

The scientific fascination with the Northern Lights has spawned a long tradition of cross-cultural cooperation and exchange. Scientists from many nations have gathered to investigate and comprehend this marvel of nature better. Not only has the exchange of information and study results improved our comprehension of auroras, but it has also promoted mutual respect and collaboration amongst different cultures.

Perspectives on Astronomy and Astrophysics:

The Northern Lights started to provide insights into other areas of astrophysics and planetary science as scientific understanding grew. Scholars have established analogies between the magnetic fields of other celestial bodies and Earth's auroras, providing insight into the intricate interrelationships among planets, stars and their surroundings.

The Digital Era and International Recognition:

The way we view and communicate about the Northern Lights has changed with the arrival of the digital era. People from all around the world can share in real time their personal experiences of seeing auroras thanks to social media and the internet. The prestige of the Northern Lights as a phenomenon and source of attraction for people worldwide has increased because to these internet linkages.

Outreach and Education Programs:

Public education initiatives about the Northern Lights have accelerated in recent years. Exhibitions and public education initiatives centered around the auroras are often held in observatories, research centers, and museums. These programs encourage awe and interest about the natural world in addition to scientific literacy.

Aurora Research's Prospects:

As a result of continuous scientific progress and space research, the study of the Northern Lights keeps evolving. Scientists are attempting to understand the complexities of the auroras' genesis and behavior as more information about them is being provided by satellites and sophisticated devices.

In conclusion, the history, culture, and science of humanity have all been profoundly impacted by the Northern Lights. The Northern Lights act as a link between our planet Earth and the heavenly wonders of the universe, inspiring everything from modern art to mythology and beliefs to scientific research and the worldwide community of aurora fans. They invite us to look up in wonder and ponder the majesty of the cosmos, serving as a constant reminder of the deep and enduring link that exists between humanity and the wonders of the universe.

CHAPTER 3

Geographic Locations for Northern Lights Viewing

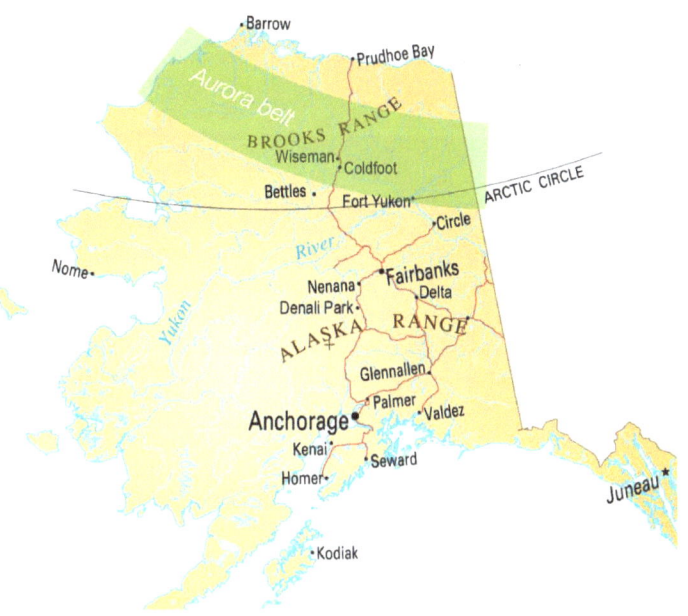

High-latitude locations, mostly in the Northern;

Hemisphere, are home to the Aurora Borealis, or Northern Lights. The Northern Lights are most frequently seen in a range of nations and areas that are well-known for their breathtaking auroral displays. We examine these locations and the distinctive sensations they provide to those who are interested in auroras below.

Scandinavia-Sweden, Finland, Norway:

One of the best places to see the Northern Lights is Scandinavia. The northern parts of Norway, which include the Lofoten Islands and Tromsø, are great places to see auroras. The northernmost regions of Sweden, including Abisko, are also good places to see the lights because of the clean, dark skies. Lapland, in Finland, especially the areas of Rovaniemi and Inari, is renowned for its regular aurora displays and charming Arctic atmosphere. For those who want to pursue the Northern Lights, a plethora of lodges and tour companies operates in these nations.

Iceland:

The "Land of Fire and Ice," Iceland, is renowned for its breathtaking scenery and incredible opportunities to see the Northern Lights. Although the capital city of Reykjavik is frequently visible from outside, more isolated areas, like

Thingvellir National Park, provide better viewing conditions and darker skies.

Canada {Yukon, Northwest Territories and Nunavut}:

The northern regions of Canada are excellent places to see the Northern Lights. Aurora fans are drawn to places like Yellowknife in the Northwest Territories and Whitehorse in the Yukon. The northern latitude, low levels of light pollution, and clear winter sky are advantageous for these areas. Another great option is the isolated village of Baker Lake in Nunavut.

Alaska (USA):

The immense wildness of Alaska offers many chances to see the Northern Lights. Because of its position beneath the "Auroral Oval," where auroras are regularly seen, Fairbanks is a popular site for aurora viewing. A well-liked location for unwinding in natural hot springs while taking in the dancing lights over head is Chena Hot Springs Resort.

Scotland (Shetland islands and Orkney}:

The Orkney and Shetland Islands in Scotland provide the opportunity to see the Northern Lights, albeit being less frequently linked to this phenomenon than some other locations. These islands are located so far north that they occasionally witness amazing light displays during times of intense solar activity.

Greenland:

The Arctic wilderness of Greenland provides an isolated and unspoiled environment for viewing auroras. The Northern Lights can be seen in a breathtaking Arctic setting in towns like Kangerlussuaq and Ilulissat.

Scotland {Cairngorms National Park}:

Another place in Scotland where one might occasionally see the Northern Lights is Cairngorms National Park. Travelers based in the United Kingdom have an easier time reaching this location.

Remote Arctic experience

Isolated Arctic trips, like those to Svalbard, Franz Josef Land, or Canada's isolated Arctic islands, provide chances to see the Northern Lights in pristine, less- traveled areas for those looking for a more daring travel experience.

Lapland: Norway, Sweden, and Finland

Lapland, which is located in Finland, Sweden, and Norway, is a fantastic place for anyone who want to see the Northern Lights. This isolated area makes for an enthralling backdrop for seeing auroras, with its snow-covered landscapes and pure woodlands. Popular sites in Lapland are Tromsø (Norway), Kiruna (Sweden), and Abisko (Sweden), which all have well- developed infrastructure and excursions catered to aurora chasers.

Scotland's Northwest:

The Scottish Highlands are another popular place to see the Northern Lights, particularly in the northwest areas of Gairloch, Ullapool, and Lochinver. Due to their relative isolation, these locations enjoy darker skies that are perfect for viewing auroras.

The Faroe Islands:

The Danish autonomous area of the Faroe Islands is well-known for its striking scenery and sporadic displays of the Northern Lights. Even though they are less frequent than in some other places, aurora sightings here are very remarkable because to the distinctive blend of untamed landscapes and ocean views.

Shetland Islands, Scotland:

Another area in Scotland that sometimes sees the Northern Lights is the Shetland Islands. The Shetlands, which are nearer Scotland's mainland than the Orkney Islands, provide visitors with the chance to see auroras, particularly when solar activity is at its highest.

Canada's Inuvik and Tuktoyaktuk:

Inuvik and Tuktoyaktuk in Canada's Northwest Territories are becoming more and more well-known for their Northern Lights encounters. These Mackenzie Delta communities offer a distinctive Arctic environment for aurora watching.

Norway's Alta:

One of Norway's top spots to see the Northern Lights is Alta, which is situated in the country's most northern region. The town is well-equipped to accommodate tourists looking for an incredible aurora experience, and it is home to the first Northern Lights observatory in history.

Arkhangelsk and Murmansk in Northern Russia:

The North Russian districts of Murmansk and Arkhangelsk provide the best chances to see the Northern Lights. Popular among those who love seeing auroras, Murmansk, which is close to the Kola Peninsula, is home to tour companies that can help you make the most of the experience.

The East Coast of Greenland:

The vast and isolated east coast of Greenland is an amazing place to see the Northern Lights. The villages of Ittoqqortoormiit and Tasiilaq offer access to unspoiled Arctic environments, often displaying auroras in the night sky.

Isle of Man (British Isles):

The island of Man, in the Irish Sea, is not a common place to see the Northern Lights, although it does occasionally have aurora sightings. The island is a fascinating choice for British stargazers and aurora fans because of its gloomy skies.

Research Stations for Astronomy and the Arctic:

There are often excellent chances to see the Northern Lights at research stations and observatories in the Arctic and Antarctic. Both for their research and their sporadic aurora displays, places like the Swedish Institute of Space Physics (IRF) research station in Kiruna, Sweden, and the

Amundsen-Scott South Pole Station in Antarctica are well known.

The Canadian Great Slave Lake:

The deepest lake in North America, Great Slave Lake is situated in Canada's Northwest Territories and provides excellent chances to see the Northern Lights. The city of Yellowknife, which is located on the lake's northern coast, is well known for its brilliant auroras and bright skies. In the winter, you can even see the lights from a warm cabin on the frozen lake.

Trips for Aurora Hunting:

The popularity of aurora hunting excursions has increased recently. These guided trips take you to the best spots— often in secluded, gorgeous settings—where you can see the Northern Lights. These adventures are made to increase your chances of seeing the breathtaking show, whether you decide to explore the Canadian wilderness, the Russian Arctic, or Scandinavian Lapland.

Places to Go Off the Grid:

Go to lesser-known, off-the-beaten-path locations in Northern Europe and North America if you'd rather get away from the crowds and have a more private experience. The charm of isolation and wild natural beauty can be found in these locations, even if they might not have as many facilities or tour guides. People from all backgrounds are captivated by the Northern Lights, a natural wonder that cuts over boundaries and cultural boundaries. Seeing the auroras is an experience that will last a lifetime, regardless of whether you go to a well-known location or explore less-explored areas. To maximize your chances of witnessing this captivating celestial occurrence, make sure to thoroughly organize your trip, confer with local experts, and monitor solar activity and weather forecasts.

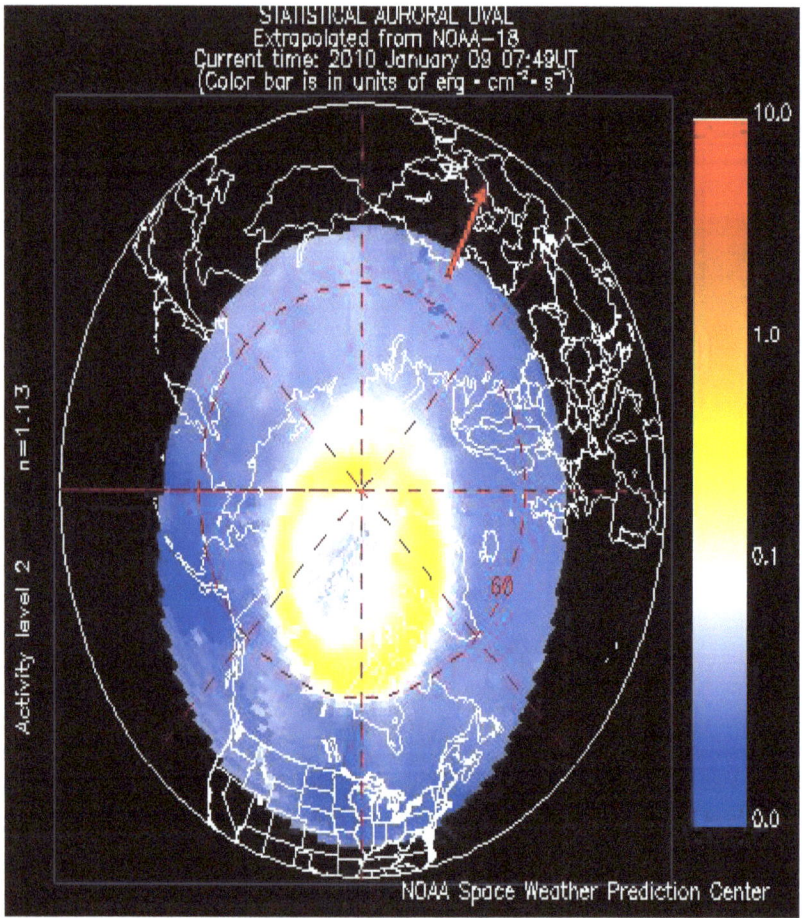

CHAPTER 4

Capturing the Magic of the Northern Lights

For individuals who are captivated by the Northern Lights, pursuing the aurora is an exciting activity. You'll be engrossed in the ethereal dance of lights in the night sky as you set out on this trip, but patience and preparation are needed to see this amazing natural show and take beautiful pictures. Let's go over the best ways to see and capture the Northern Lights for a truly unique experience.

Time and Place:

Pick the right season: Generally speaking, the best time of year to see the Northern Lights is from September to April, when the evenings are the longest. Still, the optimal timing to see them will depend on where you are.

Choose the ideal spot: Do your homework and look for a place where there is a good chance of seeing aurora borealis. High-latitude destinations for aurora viewing, such Norway, Canada, and Sweden, are well-known for their beautiful sky and minimal light pollution.

Forecast for Aurora:

Stay informed: Keep an eye on solar wind, geomagnetic conditions, and aurora activity in real time by using aurora forecast websites and apps. Important information can be found on websites like the Aurora Forecast and the Space Weather Prediction Center.

Weather Conditions:

Look for clear sky: For the best viewing and photography, clear, dark skies are essential. Keep an eye on weather forecasts to make sure the conditions are ideal.

Camera Hardware:

Use a tripod: To get clear, long-exposure photos of the Northern Lights, a sturdy tripod is a need.

Select the appropriate lens: To capture the whole aurora show and let more light into the camera sensor, a wide angle lens with a large aperture (such as f/2.8 or lower) is preferable.

Modify the camera's settings: Switch to manual mode and play about with the exposure parameters. Start with a wide aperture, an ISO of 800-1600, and a shutter speed of 10–30 seconds. When capturing long-exposure photos, think about using a self-timer or a remote shutter release to minimize camera vibration.

Elements:

Frame your shot: To give your Northern Lights photos more depth and perspective, arrange your shots with intriguing foreground features like trees, mountains, or water.

Timing and Patience:

Have patience: Aurora activity is erratic, so be ready to bide your time until the lights arrive. They show up for a long show at times, or they might only come up for a short while.

Get going early: Before the aurora is predicted to arrive, set up your gear and be prepared for action. This guarantees that you won't overlook any amazing moments you won't overlook any amazing moments.

Light Pollution: Reduce light pollution by staying away from locations that have an abundance of   artificial light. If you want to get the best shots of the Northern Lights, stay away from towns and cities.

Ensure your Comfort and Warmth:

Honor nature: Take note of your surroundings when going aurora chasing. Steer clear of delicate flora and leave no record of your visit. Respect the local populations' traditions and customs when visiting your chosen location.

Final Steps:

Edit your images: Once you've taken your pictures, use post-processing software, such Adobe Lightroom or Photoshop, to make the most of your Northern Lights photos by enhancing and fine-tuning your shots. You can do this by altering exposure, contrast, and colors.

CHAPTER 5

The Aurora Borealis Variability and Predictability

The stunning displays of vivid hues and dancing lights in the arctic night skies are known as the Aurora Borealis, or Northern Lights.

The intensity and frequency of auroral displays are governed by a complex interplay of solar activity, the solar cycle, geomagnetic conditions, and weather factors. These captivating occurrences show a surprising degree of unpredictability. A combination of natural phenomena, technical advancements, and scientific knowledge are used to forecast and follow the Northern Lights.

Sunspots and Solar Activity:

In particular, solar storms and sunspots are the main sources of solar activity that cause the Northern Lights. These include charged particles (solar wind) and magnetic disturbances coming from the Sun and are linked to increased solar radiation. The solar cycle, an 11-year cycle that consists of periods of high and low solar activity, is when the most notable fluctuations in solar activity take place.

Maximum Solar:

Maximum Solar: The Sun is most active during the solar maximum period, when there are more sunspots and solar flares. More charged particles are being sent towards Earth by this increased solar activity, which raises the possibility of strong and frequent auroral displays.

Minimum Sunlight:

On the other hand, the Sun becomes less active and there are fewer sunspots during the solar minimum phase. As a result, the frequency and intensity of the Northern Lights are diminished as fewer solar disturbances make it to Earth.

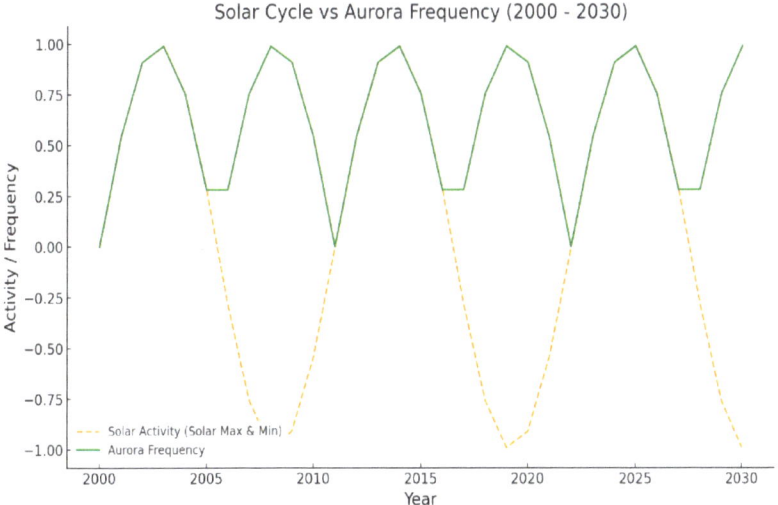

Earth's Magnetic Field:

The Northern Lights are greatly influenced by the Earth's magnetic field and how it interacts with the solar wind. Magnetic fields and energy are carried by the charged particles that make up the solar wind. When it gets to Earth, it has the potential to cause auroras and disturb the magnetosphere of the planet.

Strong Solar Storms:

Strong solar storms, known as geomagnetic storms, have the ability to greatly increase auroral activity. The Northern Lights are caused by charged particles entering the Earth's atmosphere as a result of magnetic reconnection caused by specific interactions between the solar wind and the magnetosphere.

Strong Solar Storms:

Strong solar storms, known as geomagnetic storms, have the ability to greatly increase auroral activity. The Northern Lights are caused by charged particles entering the Earth's atmosphere as a result of magnetic reconnection caused by specific interactions between the solar wind and the magnetosphere.

Season and Time of Day:

The season and time of day have an impact on how visible the Northern Lights are. The winter months in the polar areas are known as "polar night," when the evenings are the longest and there are more chances to see auroras. Though they are frequently less common, auroras can still be seen in different seasons.

Resources and Instruments for Forecasting and Monitoring Auroras:

Aurora Forecast Applications and Websites:

Real-time aurora forecasts and predictions based on solar activity and geomagnetic conditions may be found on a number of websites and smartphone apps. The Space Weather Prediction Center, Space, Weather, Live, and applications like Aurora Alerts are notable sources.

Indices of Geomagnetic:

Indexes such as the planetary A index and the Kp index are used by scientists to evaluate geomagnetic activity and its possible influence on auroras. A higher Kp index means there's a better chance of seeing auroras.

Solar Observatories:

Solar observatories, like the Solar and Heliospheric Observatory (SOHO) and the Solar Dynamics Observatory (SDO), track the

Sun's activity and assist in forecasting solar storms and eruptions that may intensify auroral displays.

Geomagnetic Observatories:

An invaluable resource for aurora forecasting, geomagnetic observatories worldwide track the Earth's magnetic field and give real-time data on geomagnetic conditions.

Astronomical Observatories:

Space telescopes such as the Hubble Space Telescope, along with astronomical observatories, help track and study solar activity and its effects on Earth's atmosphere.

Guides and Tour Operators in the Area:

Local tour operators and guides frequently have first-hand knowledge of the weather and the ideal times and locations to watch the Northern Lights while visiting well-known aurora spots.

Community Engagement:

By interacting with the community of people who chase the Northern Lights on social media and in online forums, you may get up-to-date information and advice on the best times and locations to see them. These tools and resources assist enthusiasts and researchers in making better informed decisions and maximizing their chances of witnessing the enchanting beauty of the Aurora Borealis, even though precisely predicting the timing and intensity of the Northern Lights remains a scientific challenge due to the numerous variables involved.

CHAPTER 6

Distinctive Elements of the Northern Lights

The Aurora Borealis, often known as the Northern Lights, is a breathtaking natural phenomenon that is well-known for its distinct and alluring colors, shapes, and patterns. The intricate interaction between charged particles from the Sun and the magnetic field of Earth produces these features.

Variety of Colors:

While there is a wide spectrum of colors that the Northern Lights can display, the most prevalent ones are green, pink, and purple. Different atmospheric gases reacting with charged particles produce these hues. Among the hues are:

Green:

At lower altitudes (about 100 kilometers or 62 miles above the Earth's surface), charged particles clash with oxygen to form green, the most prevalent color of auroras.

Red:

 Occurs less frequently and is linked to oxygen collisions at higher altitudes. Red and green are a common combo.

Purple and Pink:

A combination of nitrogen and oxygen reactions at different elevations produce these hues. Blue: happens when charged particles excite nitrogen molecules.

White and Yellow:

These hues are uncommon and usually appear when nitrogen is present.

Forms and Designs:

Auroras can appear in a range of patterns and shapes that shift and vary with time. Among the prevalent forms are: Curtains: One of the most identifiable auroral shapes are vertical curtains of light, which resemble hanging drapes in the sky.

Corona:

A diffuse, round glow with outward-extending rays is what is seen above during a corona.

Bands and Arcs:

These are sky formations that are horizontal in shape and frequently resemble bright ribbons of light.

Rays:

Rays are tiny beams that resemble sunlight that emerge from the horizon.

Pulsating Auroras:

These are distinguished by abrupt shifts in intensity and frequently provide a flickering impression.

Diffuse Auroras:

Compared to other shapes, these are less structured because they cast a soft, diffused illumination across wide portions of the sky.

Auroras Crowns:

These have the form of an above crown with outward-extending rays, giving the impression of a regal structure.

Scientific Explanations:

The many gases in Earth's atmosphere and the altitudes at which they interact with charged particles can be used to explain the patterns, colors, and shapes of the Northern Lights. When stimulated, various gases release distinct colors, and the shapes and patterns that are seen depend on the altitude at which these interactions take place.

Northern Lights Cultural Significance:

Communities living in high-latitude areas have seen tremendous cultural and religious influence from the Northern Lights throughout history. These hypnotic lights have inspired creativity and sparked myths.

Indigenous Cultures:

Inuit: The Northern Lights are viewed in Inuit tradition as the dancing ghosts of the dead. They are thought to assist deceased people's spirits in transitioning to the afterlife. Sami: The Sami people of the Arctic region consider the Northern Lights to be a sign of connection with the spirit realm and think they are the souls of the dead.

The Norse Tradition:

According to Norse mythology, the Northern Lights were connected to the Valkyries' armor—warrior maidens who selected fighters deserving of entry into Valhalla. The lights were interpreted as a celestial manifestation of God's judgment and presence

Perceptions in China:

The Northern Lights were called "flames in the sky" or "banners" in old Chinese writings. It was common to interpret the lights' presence as an omen of impending important occasions, both fortunate and unfortunate.

Literature and Folklore:

Around the world, literature, art, and mythology have all been profoundly impacted by the Northern Lights. Numerous tales and poetry have been written about them in an effort to convey the wonder and amazement they evoke.

Contemporary Cultural Importance:

Many cultures around the world still honor and celebrate the Northern Lights today. In areas like Scandinavia and Canada where they are frequently seen, the lights serve as a symbol of pride and identity for the country. The beauty and cultural significance of the Northern Lights are celebrated through festivals and celebrations.

Quotes from Enthusiasts:

"It feels like a living, breathing thing, the Northern Lights." They begin softly and then explode into a vibrant, colorful dance. It resembles seeing enchantment in the heavens." Sarah from Norway.

When I first saw the Northern Lights, I had the impression that I was a part of something far greater than myself. It's a deeply moving and humble experience." Alex from Canada.

"In the heart of winter, when the night is at its darkest, the Northern Lights are a reminder that even in the coldest and darkest moments, there is beauty and wonder in the world." Elena from Finland.

The distinct hues, forms, and patterns of the Northern Lights have had a profound impact on human culture and consciousness. They keep us connected to the secrets of the cosmos and the eternal force of natural beauty, and they continue to be a source of wonder, inspiration, and spiritual significance. The mysteries of the cosmos and the enduring power of natural beauty.

CHAPTER 7

Aurora Borealis Scientific Research and Its Contributions

Space Weather Forecasting:

To enhance space weather forecasting, scientists constantly track solar activity and geomagnetic conditions. In particular, protecting satellites, electrical grids, and communication networks depends on our ability to comprehend the mechanisms that cause solar storms and their possible effects on Earth.

Physics of the Magnetosphere:

Our knowledge of the Earth's magnetosphere—the area beyond our planet that is impacted by its magnetic field— is enriched by ongoing research on the Northern Lights. The interaction between the solar wind and Earth's magnetic field is studied by scientists using data gathered during auroral displays. This helps us understand how the magnetosphere shields our planet from dangerous cosmic radiation.

Interactions Between the Sun and Earth: The dynamic link between the Sun and Earth is investigated in this field of study. It seeks to understand the processes that cause charged particles to be released from the Sun and how they affect the atmosphere and magnetic field of Earth. This work sheds important light on how solar events might interfere with electricity grids, radio communications, and navigation systems.

High-Altitude Observations: To gather data from the auroral zone, scientists use cutting-edge equipment and high-altitude research platforms, like rockets and high-altitude balloons. These observations provide information on the chemical reactions and energy transfer involved in auroral displays, allowing researchers to better understand the upper atmosphere and the processes that contribute to them.

Remote Sensing and Imaging:

Real-time monitoring of the Northern Lights is made possible by remote sensing techniques, which include the use of satellite- and ground-based equipment. Scientists evaluate how solar activity affects Earth's atmosphere and develop prediction models using data from this equipment. High-resolution imagery contributes to our understanding of the development of auroral displays by enabling the

capture of their minute details.

Mechanisms of Particle Acceleration:

Acceleration of charged particles is directly related to the Northern Lights. Scientists look into the processes that drive these particles in the magnetosphere to extremely high energy. Gaining an understanding of these processes is crucial to understanding how particles behave in space and how they affect Earth.

Space-Based Missions:

To better understand Earth's magnetosphere and its interactions with the Sun, space organizations from all over the world, including NASA and the European Space Agency (ESA), perform missions. NASA's Magnetospheric Multiscale (MMS) project, for example, investigates the basic mechanisms of magnetic reconnection, which is essential to auroral displays and space weather phenomena.

Citizen scientific Initiatives:

The public is invited to participate in citizen scientific initiatives by providing observations of auroras using platforms such as Auroras urus. Together, these observations improve our knowledge of the size and frequency of auroral displays and offer a useful dataset for future scientific study.

Continuing scientific investigations into the Aurora Borealis are essential to advancing our understanding of space weather, solar-terrestrial interactions, and the Earth's magnetosphere. The results of these research not only broaden our knowledge of the natural world but also have applications in satellite communication, space exploration, and catastrophe preparedness.

Reconnection of the Magnetosphere:

Magnetic reconnection is one of the main processes that is investigated in conjunction with the Aurora Borealis. Charged particles can enter the Earth's magnetosphere by this process, which is the connecting and rearranging of magnetic field lines from the Sun and Earth. To learn more about how this mechanism impacts the movement of particles and energy between the Sun and Earth, researchers examine the specifics of this process during auroral phenomena.

Effect on Communication and Satellite Systems

The possible effects of solar activity and the ensuing Northern Lights on satellite communication and navigation systems are the subject of ongoing research. It is crucial to comprehend the connection between interruptions in these technologies and space weather events in order to create mitigation plans and protect vital infrastructure

Chemical composition and spectral analysis

Spectroscopy is used by researchers to examine the spectral features of auroral emissions. Scientists can determine the chemical makeup of the upper atmosphere and learn more about how atoms and molecules behave during auroral displays by analyzing the light wavelengths emitted during: these phenomena.

Effects of Ionosphere:

Ionospheric interactions are responsible for the Northern Lights, and scientists study how these interactions impact the ionosphere's activity. Gaining an understanding of these phenomena is crucial to enhancing the precision of GPS and other radio wave propagation and navigation systems.

Coupling of Atmospheric Layers:

Research on the Northern Lights advances our knowledge of the intricate interactions between the thermosphere, mesosphere, and exosphere, three of Earth's atmospheric layers. Research on the energy and momentum transmission between these layers during auroral phenomena is still ongoing.

Outreach and Education:

Research on the Aurora Borealis includes outreach and educational initiatives meant to increase public knowledge of and enthusiasm for magnetospheric physics and space weather. These programs raise public awareness of the importance of studying the Northern Lights and serve as an inspiration for the next generation of scientists.

Cross-border Cooperation:

International cooperation is essential because of the global character of the Northern Lights and their relationship to solar activity. International collaboration among researchers allows for the exchange of information, expertise, and materials, enabling a more thorough and well-organized investigation of auroral occurrences. Our dedication to solving the universe's secrets and shielding our planet from space weather is demonstrated by the ongoing scientific study of the Aurora Borealis. It offers applications for space exploration, climate monitoring, and disaster preparedness in addition to expanding our knowledge of Earth's magnetosphere and solar-terrestrial interactions.

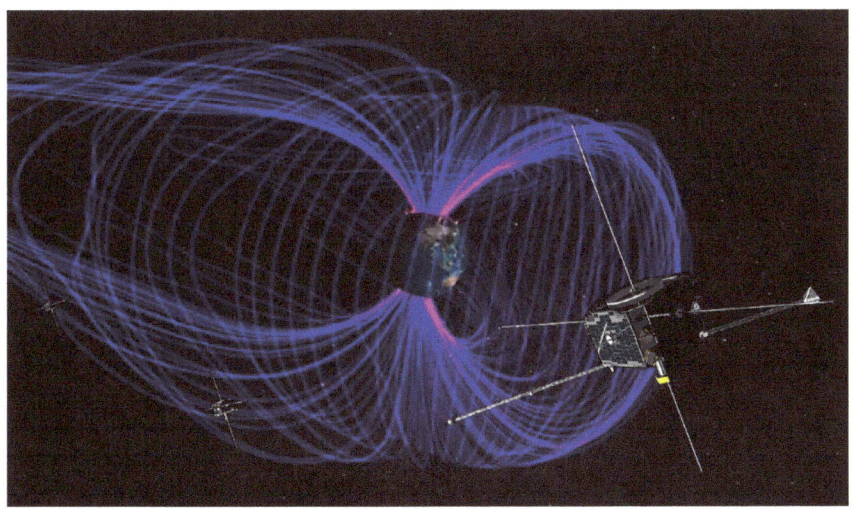

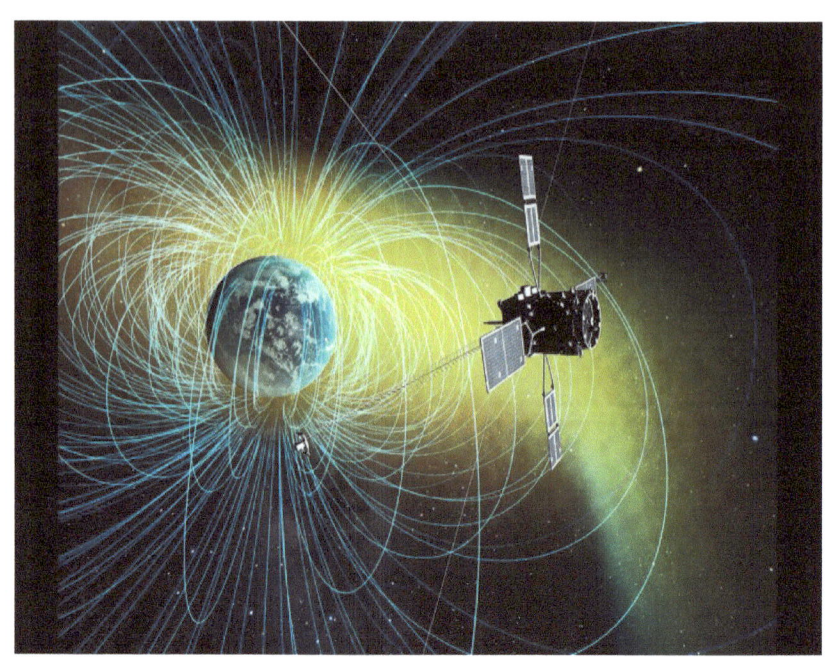

NASA's THEMIS mission satellites observing auroral activity in Earth's magnetosphere.

CHAPTER 8

Environmental Effects of Climate Change and Light Pollution on the Northern Lights

In addition to being a natural wonder, the Aurora Borealis, commonly known as the Northern Lights, is a celestial spectacle that is susceptible to numerous environmental influences such as light pollution and climate change. It's critical to recognize the dangers these breathtaking shows face and the significance of preserving the clear skies that support the Northern Lights as we work to conserve them for future generations.

Light-related pollution:

Intrusion of Artificial Light:
The Northern Lights can be obscured by light pollution from cities. In addition to reducing visibility, the brightness of city lights

also makes the auroras less spectacular. It is frequently difficult for onlookers in metropolitan environments to fully appreciate the beauty of this natural occurrence.

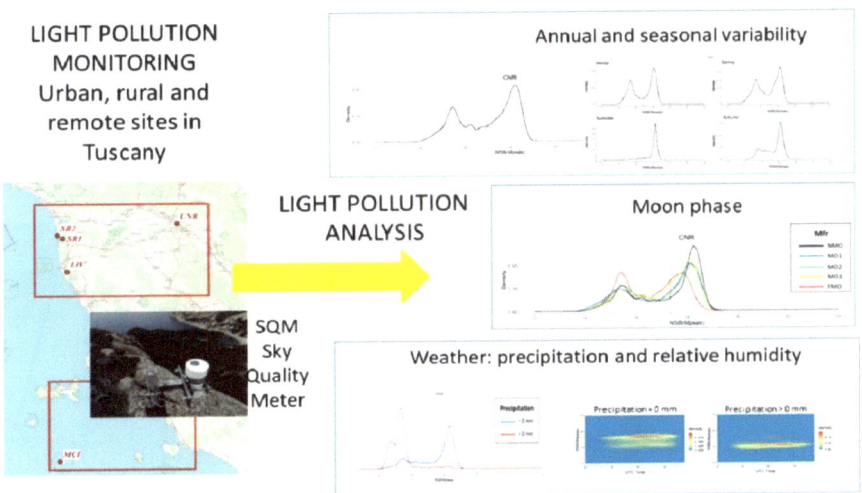

Diminishment of Ambient Darkness:

Overindulgence in artificial nighttime illumination diminishes the amount of natural darkness needed to see the Northern Lights. People living in urban areas may lose sight of the wonders of the night sky in locations where light pollution is severe enough to make lights nearly invisible.

Global Warming:

Effect on Solar Activity:

Climate change has the potential to impact solar activity, which in turn can impact the frequency and power of the Northern Lights. The solar wind and Earth's magnetic interactions can be disrupted by changes in the Sun's behavior, which may change the frequency and patterns of auroral displays.

Change in Seasons Viewed:

Seasons and the timing of the polar night are further aspects of climate change. The opportunity for seeing the Northern Lights may shift as winters get shorter or have milder temperatures. The tourism sector in areas where aurora-related activities are important may be impacted by this.

Why It's Important to Protect Dark Skies:

Scientific and Cultural Significance:

Both scientifically and culturally, the Northern Lights are of great significance. They continue to further our knowledge of space weather and Earth's magnetosphere while serving as the inspiration for innumerable myths, tales, and artistic creations. In order for future generations to engage with this celestial splendor and understand its cultural and scientific value, dark sky preservation is necessary.

Local economies and tourism:

An important source of revenue for many areas is aurora tourism. Communities that depend on this natural attraction may suffer economically if light pollution and climate change cause the Northern Lights to become less visible and appealing.

Equilibrium Environment:

Keeping the skies dark helps the environment as well as people's enjoyment of the night sky. Numerous nocturnal creatures and flora depend on the cycles of natural light. These cycles are upset by light pollution, which can also have negative impacts on the ecosystem and wildlife.

Educating the Public:

Maintaining dark skies and the ability to see the Northern Lights helps raise public awareness of more general environmental issues, such as the significance of combating climate change and minimizing light pollution. It motivates people to take steps in favor of a sustainable and ecologically sensitive way of living.

Keeping the dark skies safe from light pollution and climate change is a common duty. Urban surroundings can become more aurora-friendly by making improvements to outdoor lighting practices, such as shielding and adopting warm-colored LEDs, to counteract light pollution. Furthermore, mitigating climate change and lowering our carbon footprint can contribute to preserving the environmental factors that are essential for the Northern Lights to thrive.

Safeguarding the cultural, scientific, and economic significance of the Northern Lights is just as important as maintaining its natural beauty. We can guarantee that next generations will be captivated by the mesmerizing dance of the Aurora Borealis in the pitch-black heavens overhead by taking action to mitigate light pollution and tackle climate change.

CHAPTER 9

Aurora Chasing Travel Tips

There are a number of useful suggestions to take into account when organizing an adventure to chase the Aurora, ranging from the ideal times to visit and locations to packing lists. Prioritizing your health and safety is very essential before setting off on this fascinating adventure.

Best Times to Go:

1. The Winter Season: Wintertime, when the evenings are longest and the skies are darkest, is when people most frequently see the Northern Lights. In high-latitude areas, September through March is the best period to see them.

2. New Moon Phase: To ensure the darkest skies and the finest aurora viewing, schedule your trip during the new moon phase.

3. Clear Nights: As cloud cover can obstruct your view of the Northern Lights, check your local weather forecast for clear nights.

Appropriate Sites:

- Aurora Belt: Pay particular attention to areas that are part of or adjacent to the "Aurora Belt," including parts of Russia, Lapland (Finland, Sweden, and Norway), Iceland, Canada (Yukon, Northwest Territories), and Alaska (USA).
- Dark Skies: Look for places like national parks, isolated wilderness areas, and designated dark sky preserves that have less light pollution.

Items to Bring:

- Wear Warm Clothes: Wear layers of clothing to stay warm in the winter. Remember to include waterproof boots, insulated coats, thermal underwear, and high-quality hats and gloves.

- Handheld: For reliable long-exposure photography of the Northern Lights, a robust tripod is a must.

- Equipment for Cameras: To properly catch the auroras, bring a DSLR or mirrorless camera with a wide-angle lens. Additional memory cards and camera batteries are also essential.

- Headlamp or Flashlight: When arranging equipment and getting around in the dark, a hands-free light source comes very handy.

- Portable Charger: Always keep your electronics charged, particularly when you're in an isolated area with possibly scarce power sources.

- Water and Snacks: Since you might have to wait outside for a while, bring some food and a water bottle.

- Maps and GPS: To travel distant places, make sure you have dependable maps or a GPS device.

- Aurora Apps: Get real-time notifications and predictions by downloading apps such as "Aurora Alerts".

Health and Safety:

- Remain Warm: In places where people can see the Northern Lights, it is often very chilly. To avoid frostbite, wear insulated boots, layer your clothing, and minimize areas of exposed skin.

- Know the Weather: Recognize the current local weather and the potential for abrupt temperature swings. Be prepared for erratic weather and make the appropriate preparations.

- Travel in Groups: Traveling in groups promotes safety. If at all possible, travel in groups. Tell someone what you're doing and when you expect to return.

- Emergency Kit: Always keep a basic emergency kit on you, filled with supplies like additional food, emergency blankets, and a first aid kit.

- Take Care of Your Step: Uneven terrain can be dangerous in the dark. To light your path, use a headlamp or flashlight, and pay attention to your footing.

- Consider the surroundings: Take in the surroundings when you see the Northern Lights. Leave no trace, pack out all rubbish, and stick to approved pathways.

- Avoid Light Disturbances: Reduce the amount of artificial light sources, such flashlights and cellphones, to avoid affecting other people's night vision and your own.

- Adhere to local guidelines: Specific regulations, such as prohibited zones, apply in some places when it comes to watching auroras. Acquaint oneself with and abide by local laws.

- Remain Updated: Watch the news and updates about the weather and road conditions in your area. Keep yourself aware about possible risks.

It's a rewarding experience to chase the Northern Lights, but you must be well-prepared for the frequently difficult conditions found in arctic locations. You can have a captivating experience and be safe and comfortable by paying attention to your health and safety and by heeding these travel advice articles.

frequently difficult conditions found in arctic locations. You can have a captivating experience and be safe and comfortable by paying attention to your health and safety and by heeding these travel advice articles.

Dear Reader,

Thank you so much for purchasing this book! Your thoughts and feedback mean the world to us. If you enjoyed reading it or found it helpful, we would greatly appreciate it if you could leave a review in the comment section. Your words not only inspire us but also help others discover the book. Thank you for your support!

www.ingramcontent.com/pod-product-compliance
Lightning Source LLC
Chambersburg PA
CBHW040321010626
45792CB00024B/2083